INSIDE MODERN GENETICS

INSIDE EVOLUTION

ROSIE BANKS

ROSEN PUBLISHING
NEW YORK

Published in 2022 by The Rosen Publishing Group, Inc.
29 East 21st Street, New York, NY 10010

First Edition

Designer: Rachel Rising
Editor: Therese Shea

Portions of this work were originally authored by Kristi Lew and published as *Evolution: The Adaptation and Survival of Species*. All new material in this edition authored by Rosie Banks.

Library of Congress Cataloging-in-Publication Data

Names: Banks, Rosie 1978- author.
Title: Inside evolution / Rosie Banks.
Description: New York : Rosen Publishing, [2022] | Series: Inside modern genetics | Includes index.
Identifiers: LCCN 2021029096 (print) | LCCN 2021029097 (ebook) | ISBN 9781499470413 (library binding) | ISBN 9781499470406 (paperback) | ISBN 9781499470420 (ebook)
Subjects: LCSH: Evolution--History--Juvenile literature.
Classification: LCC QH367.1 .B36 2022 (print) | LCC QH367.1 (ebook) | DDC 576.809--dc23
LC record available at https://lccn.loc.gov/2021029096
LC ebook record available at https://lccn.loc.gov/2021029097

Manufactured in the United States of America

CPSIA Compliance Information: Batch #CWRYA22. For further information, contact Rosen Publishing, New York, New York, at 1-800-237-9932.

CONTENTS

INTRODUCTION

Imagine you are at your family reunion. You tell everyone to gather together for a photo. Besides the people in your immediate family, a giraffe, crocodile, eagle, and honeybee crowd into the picture. That might sound weird, or at least like a funny children's storybook, but in a way, it is not so far from the truth, at least from a scientific point of view. All these creatures are related. They have a common ancestor. In fact, all living things share an ancestor, not just animals. That means any organism you can think of—bacterium, fruit, fungus, vegetable, you name it!

Earth supports an incredible variety of life. Some are just a single cell. These are the eukaryotes, bacteria, and archaea. Then, there are animals—including people—that have trillions of cells! And life on Earth is everywhere, including Antarctica, the ocean floor, and the deepest, darkest caverns. However, we also know that not all kinds of life-forms have survived over time. Perhaps the most famous examples are the dinosaurs. Woolly mammoths and saber-toothed cats are other well-known extinct species.

In fact, about 250 million years ago, scientists estimate that more than 90 percent of all the animals that lived on Earth became extinct. Why did so many disappear? These species could not change, or adapt, to new environments quickly enough to survive. Today, many types of animals, from the gray wolf to the blue whale, are in danger of going the way of the dinosaurs.

Dinosaurs roamed Earth for 160 million years. They died out about 65.5 million years ago in what is known as the Cretaceous-Tertiary extinction event.

Where did all the plants, animals, and other creatures come from? Most scientists agree that all the living things on Earth today originated from one common, single-celled ancestor that lived about 3.5 billion years ago. That means that all the millions of species alive today (scientists have discovered and described about 1.2 million species so far) are related to one another—and to humans.

Long before Charles Darwin published his book *On the Origin of Species by Means of Natural Selection* in 1859, people noticed and wondered about the similarities and differences between life-forms. Darwin compared the way living things connect to one another with the branches of a tree. Today, that image remains. During Darwin's time, scientists relied only on the similarities and differences that they could see to decide how species might be related to one another. Today, scientists have much more sophisticated methods. They study not only what they can see with the naked eye, such as an animal's anatomy, but also what they cannot see easily, such as an organism's genetic makeup. Advances in the science of genetics have helped scientists find connections between organisms. In turn, this has helped them develop a clearer image of how the diverse life-forms on Earth fit into the tree of life.

The idea that all life-forms originated from a common ancestor and have changed over time into the variety of life that people see today is called evolution. It is supported by many scientific studies and observations. Some people like to point out that evolution is only a theory. It is true that scientists call it a theory. However, the word "theory" in science does not have the same meaning that it does in everyday speech. In casual use, the term "theory" can mean "a guess or an opinion." In science, a theory signifies a well-studied explanation of observed and experimental data. Evolution, like other theories, has been subjected to many tests. So far, scientists have found nothing that would disprove it. Does that mean that the theory will never change? No. As with all scientific theories, the evolutionary theory is subject to change when new data becomes available. However, as of today, the theory of evolution is the best fit for the scientific evidence that biologists have observed so far. It is accepted by most scientists. What makes these scientists so willing to accept the theory of evolution? To understand their reasons, people have to first examine how living organisms are related to one another.

Evidence shows that both the physical and behavioral traits shared by all human beings originated from apelike ancestors. This evolutionary process took approximately 6 million years.

CHAPTER 1

EXAMINING INHERITANCE

At the center of evolutionary theory is the idea of a "universal common ancestor." According to this concept, all life—every species you can think of—evolved from a single ancestor. Think about how you and your cousins share a grandmother. It is a bit like that but on a much, much larger scale. Species as diverse as oak trees and earthworms share an ancestor way, way down the evolutionary trail. Obviously, some organisms are more closely related to others. Different species of roses, for example, have more in common with each other than they do with elephants. Yet, these living things and all others are still linked by an evolutionary connection.

Many things on Earth change over time. Trees, for example, lose their leaves in the fall and their leaves grow back in the spring. A tadpole hatches from an egg and eventually turns into a frog. A car abandoned in a field rusts away. People age and their looks change. However, none of these is an example of evolution. What makes evolution different is that it involves genetic changes within a population. These changes are passed down from generation to generation.

THE BASICS OF HEREDITY

Your ancestors are the people in your family who came before you, like your grandparents and great-grandparents. Each ancestor passed down certain characteristics to you. Some of these characteristics, such as hair color, eye color, and height, you can see with the naked eye. Others, such as blood type or the likelihood of developing a particular disease, you cannot see. These genetically determined characteristics are called traits.

Identical twins are a powerful example of the same or very similar genes at work in two people.

Traits are passed down through families from one generation to the next. If your parents have dark hair, for example, you are likely to have dark hair. But this is not always the case. Sometimes a trait skips from one generation to the next. For example, instead of having dark hair like your parents, you might end up having bright red hair like your great-aunt. No matter, the trait was still inherited from your ancestors. The particular combination of traits you inherited makes you unlike anyone else in the family. You are genetically unique, unless you have an identical twin. Identical twins can be genetically the same and, therefore, share many of the same traits. Scientists call the combination of observable traits that a person has a phenotype. For example, if you have red hair, green eyes, and freckles, this combination would be considered part of your phenotype.

Your phenotype is controlled by information present in almost every cell of your body. This information not only tells your body to make red hair, green eyes, and freckles, but also how to grow and function. The information is written in a code that your cells can read. It is stored in a chemical molecule called deoxyribonucleic acid, or DNA.

If you could see the DNA molecule, it would look a lot like a twisted ladder. The sides of the twisted ladder are made up of two chemicals—a sugar and a phosphate. Scientists call these chemicals DNA's sugar-phosphate backbone. The rungs of the ladder are made up of chemicals called bases. There are four bases in DNA: adenine (A), guanine (G), cytosine (C), and thymine (T). The bases are found in pairs. Adenine always pairs with thymine. And guanine pairs with cytosine. Scientists have found that there are approximately 3 billion base pairs in the human genome. A genome is all the genetic material in an organism.

All living forms use this same double-helix structure within their cells to pass genetic information on.

PASSING INFORMATION

Some DNA base pairs work together in very long groups. These groupings are called genes. The order, or sequence, of the DNA base pairs in a gene tells the body how to make chemicals called proteins. Proteins are important in determining what a person looks like and how the body functions. The 3 billion base pairs of the human genome seem to be divided into about twenty thousand genes. With the exception of some identical twins, everyone's base pair sequence is slightly different. This difference in the order of base pairs is what makes people unique. However, most of the DNA sequence is the same for all humans. In fact, scientists estimate that an individual's DNA differs from all other humans' DNA by about 0.1 percent. Furthermore, a person's DNA is about 98.8 percent identical to that of the chimpanzee, humans' closest living genetic relative in the animal world.

This is a karyotype of a human male. A karyotype is an image of an organism's chromosomes.

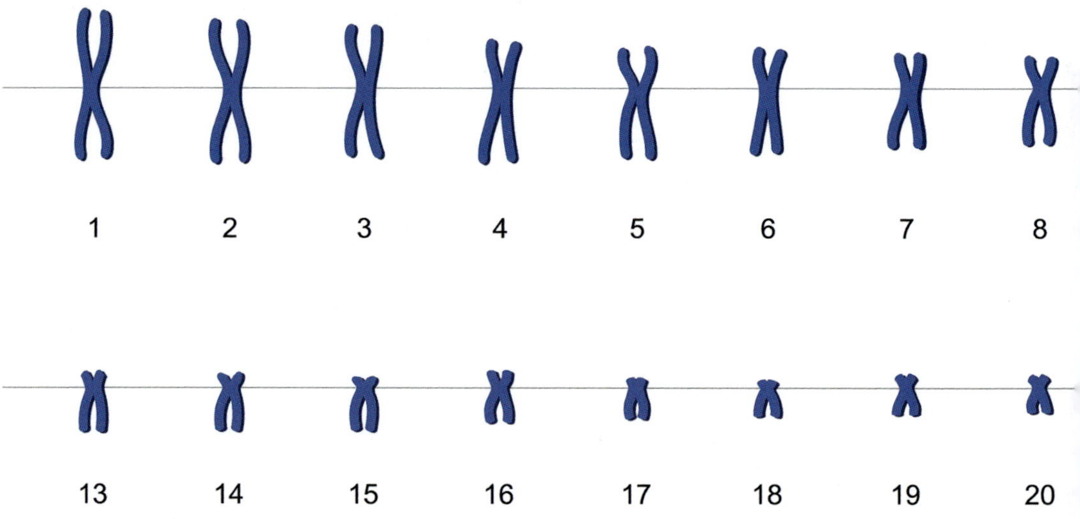

Genes are passed down from parent to child in structures called chromosomes. In most cells, humans carry forty-six chromosomes. Chromosomes come in pairs. Humans get one chromosome of each pair from their mother. The other chromosome comes from their father. Of the twenty-three pairs of chromosomes in a human, twenty-two are nearly the same size and carry similar genetic information. These chromosomes are called autosomes. The last pair is called the sex chromosomes. This pair of chromosomes determines a person's sex. Women inherit two X chromosomes while men inherit one X and one Y chromosome.

Some genes passed down from your mother and father are the same. But others are different. Different versions of the same gene are called alleles. For example, whether or not an individual has freckles is a genetic trait. There are two different freckle alleles. One tells the body to make freckles when exposed to the sun; the other tells the body not to make freckles.

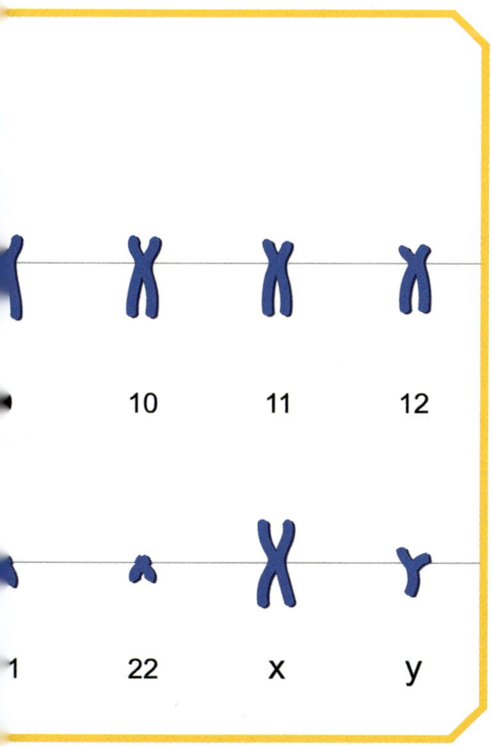

The allele that tells the body to make freckles is the dominant allele. Dominant genes will show their effect in a person's phenotype even if there is only one copy of that gene. So if either parent passes on the gene that tells the body to make freckles, their offspring will have freckles. The version of the gene that tells the body not to make freckles is called a recessive gene. In order for recessive traits to show up in the offspring, both parents must pass down the same recessive gene.

Having freckles or not is controlled by a single gene, but not all traits are controlled by a single gene. In fact, scientists think that most human traits are controlled by multiple genes. They call these traits polygenic traits. The shape of your face, the amount of curl in your hair, your IQ, and your skin color are all believed to be polygenic.

Height is another trait that is polygenic. Scientists are not sure how many genes are involved in determining a person's height.

GENE SHUFFLING AND MUTATIONS

Most cells in the human body have forty-six chromosomes in them. Ones that do not are red blood cells, which have no nucleus and no DNA, and gametes. Gametes are reproductive cells. A female gamete is called an egg. The male gamete is a sperm. In humans, these cells contain only twenty-three chromosomes. During sexual reproduction, an egg fuses with a sperm. The resulting cell has forty-six chromosomes and, if all goes well, it divides and grows to become a child.

To produce cells with half the number of chromosomes, gametes must go through a process called meiosis. Before meiosis begins, each DNA molecule is copied. As a result, instead of one twisted ladder of DNA, each chromosome contains two identical twisted ladders of DNA. (Scientists call the twisted-ladder shape a double helix.) Now the cell has twice the usual amount of DNA. During the first phase of meiosis, each chromosome pair (one from the mother and the other from the father) lines up side by side. Sometimes they exchange bits of DNA. This swapping of genetic material is called recombination, or crossing over. Recombination creates a huge opportunity for genetic variation. Genes get shuffled around on chromosomes and a large number of possible gene combinations can be produced. This gene shuffling is what causes siblings to look different, though similar. Brothers and sisters have some genes in common and these genes make them resemble one another. But they also have some genes that are different, which makes each distinct.

INVESTIGATING IDENTICAL TWINS

Identical twins form from the same fertilized egg, so they receive the same genetic material from their parents. (In contrast, fraternal twins form after the fertilization of two separate eggs.) For years, scientists believed that identical twins were genetically identical. However, studies now show that is not necessarily true. A study published in the journal *Nature Genetics* in January 2021 revealed that identical twins differ by an average of five genetic mutations. The research examined 381 pairs of identical twins and two sets of identical triplets. Most had a few points of genetic difference, but thirty-nine had more than one hundred differences in their DNA. Just thirty-eight could be considered genetically identical.

In the past, differences in genetics in twins were thought to be caused by the environments in which they grew up. This study reveals that scientists may have overestimated nonbiological factors in genetic differences between twins. In fact, twins' genomes can differ in small but significant ways from the first days of their development.

Identical twins have been invaluable to genetic studies as scientists attempt to unravel the influences of genes and the environment on certain traits. For example, if identical twins are raised in different places and one has autism while the other does not, the scientific interpretation in the past was that something in the twin's environment was the cause of his or her autism. But such an assumption can no longer be made. Scientists must determine whether a mutation in a gene might exist in one of the twins to have caused the condition. Even slight genetic tweaks at initial stages of development might be the trigger.

As a cell continues through the stages of meiosis, the chromosome pairs are pulled apart and two daughter cells are created. This division reduces the number of chromosomes in each daughter cell by half. Because of recombination, each daughter cell has a unique DNA sequence. But, at this stage, the chromosomes in the daughter cells still consist of two double helices of DNA, twice the amount of DNA needed. To reduce the amount of DNA, each daughter cell goes through another division. During this division, the two double helices of DNA that make up each chromosome are pulled apart. The result is four daughter cells with twenty-three chromosomes, each consisting of a single double helix of DNA with a one-of-a-kind sequence. During fertilization, these cells produced by meiosis from a male and female will fuse to create a cell with two copies of each chromosome, becoming a zygote.

This illustration shows the stages of meiosis, a process that occurs in all sexually reproducing organisms.

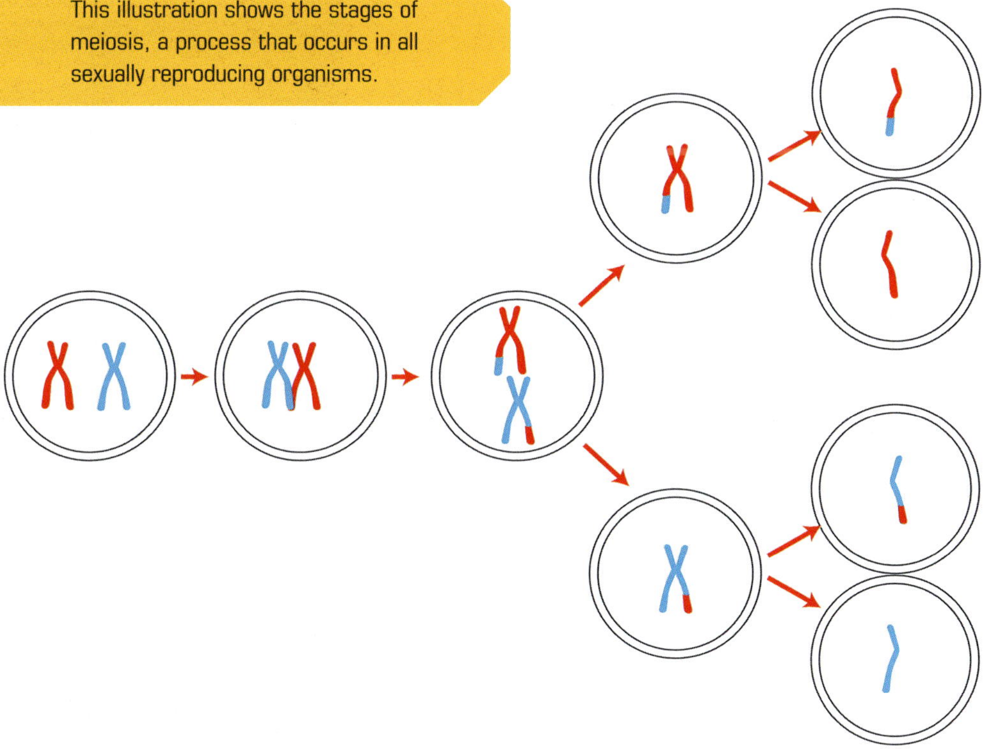

Random combination of gametes and recombination both increase genetic diversity in the offspring. However, these are not the only means of genetic variation. Mutation can cause a permanent change in DNA too. Changes in the DNA sequence may be caused by a copying error during cell division. Other changes may be caused by conditions in the environment, such as exposure to radiation or chemicals. If the DNA is damaged in a body cell, the mutation cannot be passed on to the next generation. But if the mutation occurs in a gamete, the new DNA sequence can be passed down from parent to child.

You may be wondering what all this information about DNA and mutations has to do with evolution. Well, a change in a life-form's DNA can cause changes in all aspects of its life—and in the life of any organism onto which it passes DNA. A change in frequency of a particular allele in a population of organisms over time is how a species evolves. Simply put, mutations are necessary for evolution to occur. However, our understanding of chromosomes, genes, and DNA is relatively recent and arose over many long years of scientists' speculation and research. Charles Darwin's famous theory of evolution is less than 200 years old. Many scientists before him helped pave the way, however.

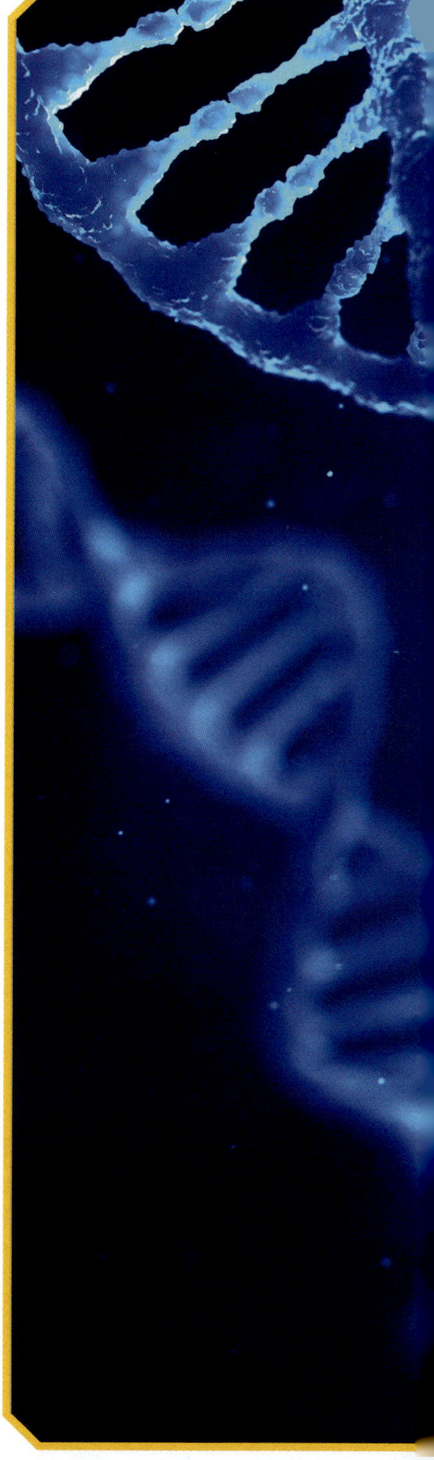

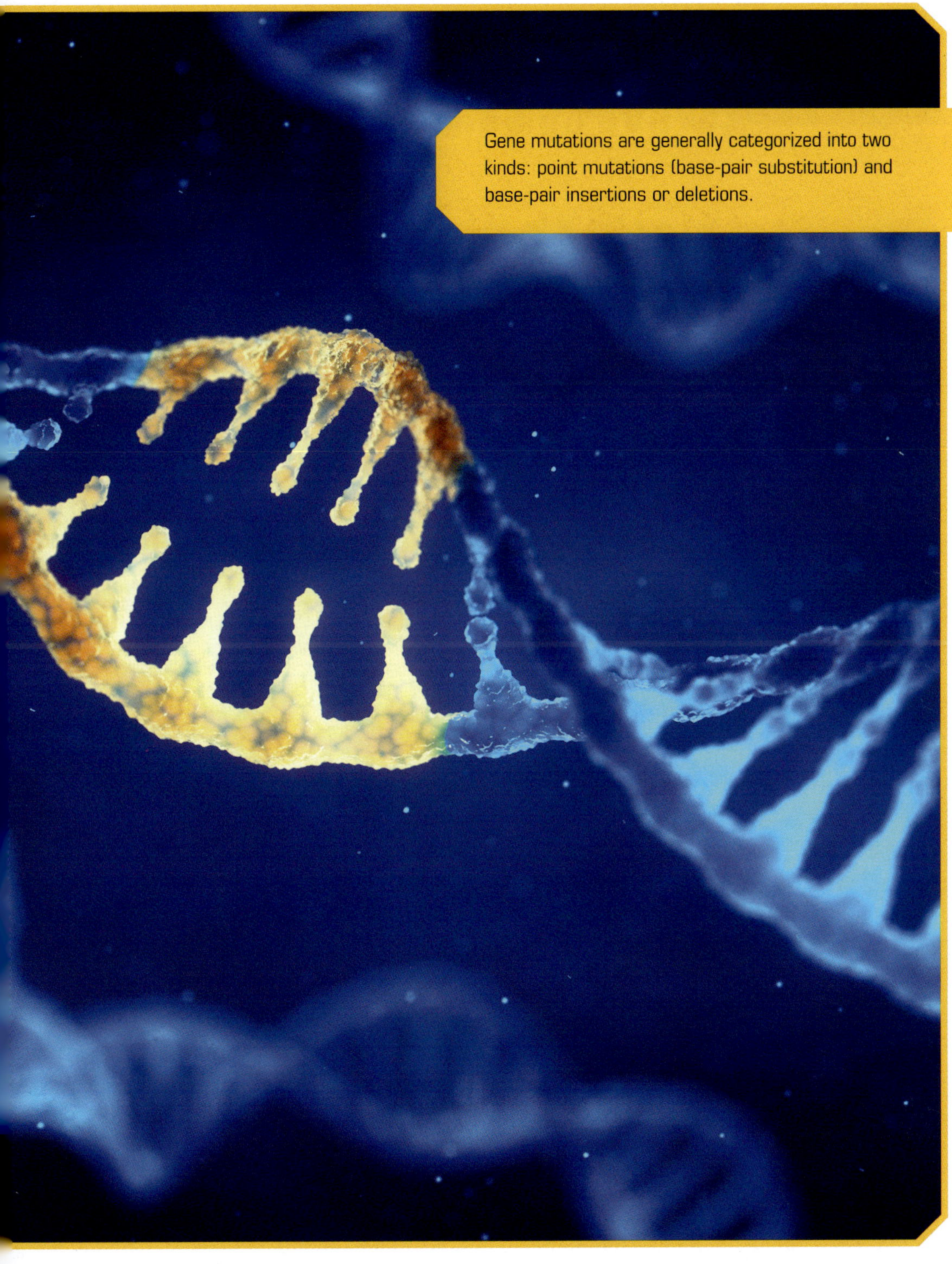

Gene mutations are generally categorized into two kinds: point mutations (base-pair substitution) and base-pair insertions or deletions.

CHAPTER 2

EVOLUTIONARY THEORY

The name Charles Darwin and the theory of evolution are closely linked and with good cause. There is no better-known scientist in the field than Darwin (1809–1882). At the time of his research, Darwin knew nothing about DNA, base pairs, and genes. The study of genetics developed long after Darwin's death. Nor did Darwin actually come up with the idea of evolution. Early Greek philosophers such as Aristotle assumed that species had arisen from a common ancestor. However, Darwin provided probable explanations about how organisms could have changed over time. His theory made sense of the natural world. Though not everyone embraced his ideas, his theory of evolution by natural selection became fundamental to modern biology.

CHARLES DARWIN

WHAT, EXACTLY, IS A SPECIES?

In the late eighteenth century, the French naturalist Georges-Louis Leclerc, comte de Buffon (1707–1788), developed the definition of the term "species" that is still used today. A species is a group of organisms that can produce fertile offspring. In other words, members of the same species can have young together and their young can also reproduce.

Different members of the same species may look very similar or quite different. Consider dogs. A Saint Bernard and a dachshund look distinct, but they are members of the same species. Scientists call their species *Canis lupus familiaris*. A Saint Bernard and a dachshund could produce puppies together. And the Saint Bernard–dachshund puppies could grow up and produce puppies of their own.

There are more than 400 recognized breeds of dogs in the world. All are the same species, however.

Some animals look like they are members of the same species, but they are not. They may even be able to produce offspring together. But they are not really the same species because the offspring they create cannot reproduce.

Horses and donkeys, for example, can reproduce. The offspring of a male donkey and a female horse is called a mule. The offspring of a female donkey and a male horse is called a hinny. However, horses and donkeys are not really the same species. Neither mules nor hinnies can have babies.

MULE

It is difficult to tell a mule and hinny apart. Scientists and breeders may use DNA testing to determine which species an animal is.

HINNY

The reason mules and hinnies are sterile has to do with the number of chromosomes in their cells. Horses have sixty-four chromosomes. Their gametes have half that number, or thirty-two. Donkeys have sixty-two chromosomes and their gametes have thirty-one. When the gametes are fused during fertilization, the resulting cell has sixty-three chromosomes. Having an uneven number of chromosomes is not a problem for the growing mule or hinny. However, when it comes to making gametes, the uneven number of chromosomes is a problem. None of the gametes that are produced has a full set of chromosomes, which is needed to form offspring.

EARLY EVOLUTIONARY WORKS

Charles Darwin was not the first Darwin to come up with the idea of evolution. In 1794 and 1796, his grandfather Erasmus Darwin (1731–1802) published a two-volume book called *Zoonomia; or, The Laws of Organic Life*. This book outlined Erasmus Darwin's ideas about the natural world. In it, he suggests that all animals could have come from one ancestor. It is considered one of the first formal theories of evolution. By studying domesticated animals and observing wildlife, he developed the idea that some animals were more likely to survive

Besides his interest in the natural world, Erasmus Darwin was also an inventor, doctor, and poet.

and reproduce than others. This idea was later expanded by his grandson, who called the process natural selection.

Another naturalist, Jean-Baptiste Pierre Antoine de Monet, chevalier de Lamarck (1744–1829) of France, published his main work on evolution, *Philosophie zoologique* ("Zoological Philosophy"), in 1809. Lamarck proposed that organisms that were placed in a different environment adjusted to their new situation. Those organisms then reproduced and passed on the new, helpful trait.

Charles Darwin would come to the same conclusion (that organisms evolved) fifty years later in his book *On the Origin of Species by Means of Natural Selection*. However, Lamarck and Darwin differed in their explanation of how this process occurred. Lamarck believed when an animal's environment changed, its needs changed. These different needs caused the animal to behave differently and use its body in different ways. Lamarck proposed that if an animal used a body structure or organ more, that structure or organ would grow larger. On the other hand, if the animal's new needs caused it to use a structure or an organ less often, the structure or organ would shrink. This is known as the use and disuse theory.

He also believed that these newly acquired traits could be passed down to the next generation. So, for example, according to Lamarck, giraffes evolved to have longer necks because their environment changed, forcing them to reach higher into trees to reach the leaves they needed to eat. Over time, Lamarck believed, the animal's neck would stretch. In addition, the giraffe's offspring would be born with a longer neck.

Lamarck also believed that evolution should be viewed more like a ladder than a tree. In other words, Lamarck believed that the goal of all creatures was to evolve into more complex life-forms. Lamarck equated more complexity with being more "perfect." He did not believe that species became extinct. Instead, they evolved into a "better" species.

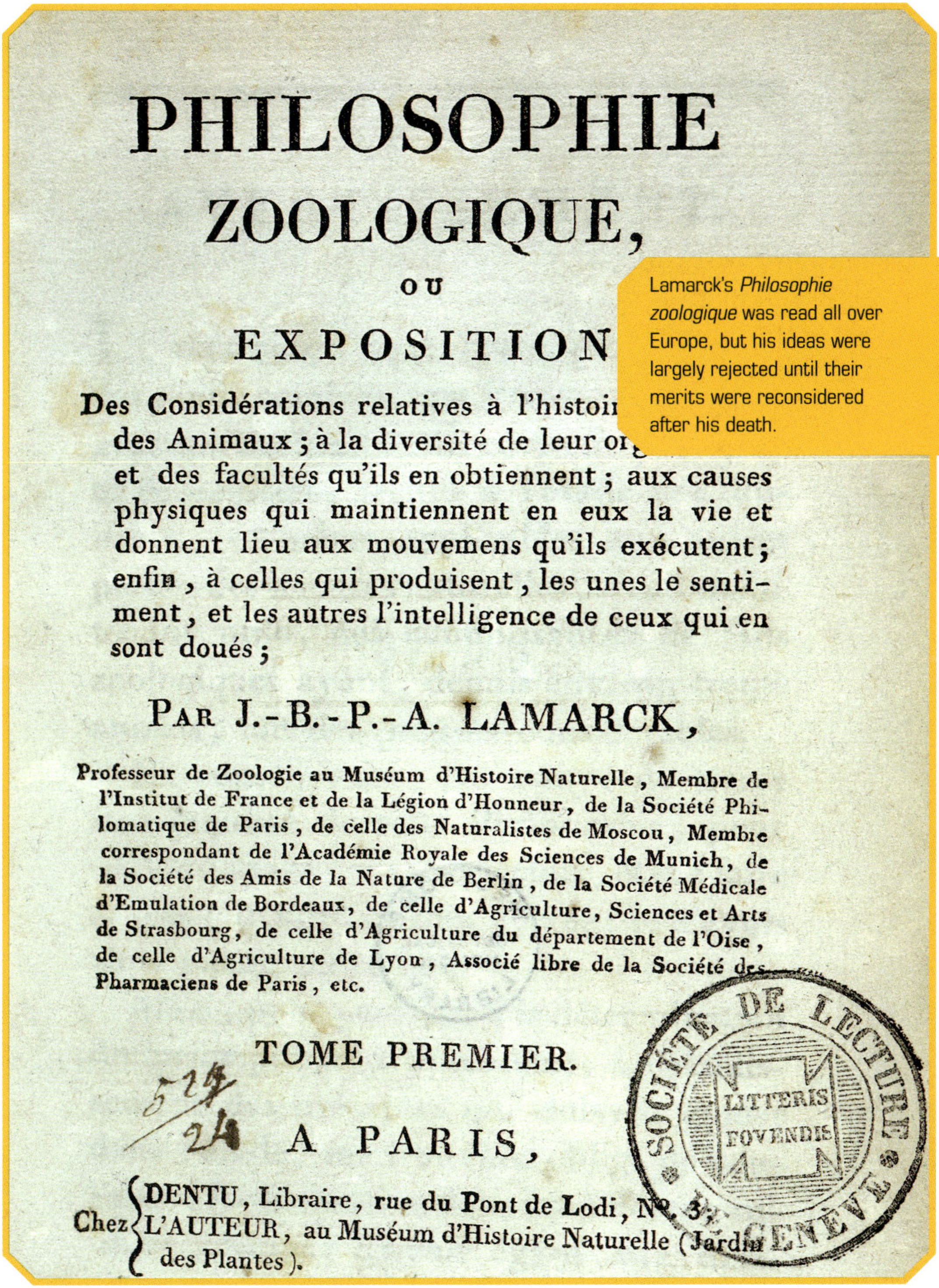

PHILOSOPHIE

ZOOLOGIQUE,

OU

EXPOSITION

Des Considérations relatives à l'histoi[...]
des Animaux ; à la diversité de leur or[...]
et des facultés qu'ils en obtiennent ; aux causes
physiques qui maintiennent en eux la vie et
donnent lieu aux mouvemens qu'ils exécutent ;
enfin, à celles qui produisent, les unes le senti-
ment, et les autres l'intelligence de ceux qui en
sont doués ;

PAR J.-B.-P.-A. LAMARCK,

Professeur de Zoologie au Muséum d'Histoire Naturelle, Membre de l'Institut de France et de la Légion d'Honneur, de la Société Philomatique de Paris, de celle des Naturalistes de Moscou, Membre correspondant de l'Académie Royale des Sciences de Munich, de la Société des Amis de la Nature de Berlin, de la Société Médicale d'Emulation de Bordeaux, de celle d'Agriculture, Sciences et Arts de Strasbourg, de celle d'Agriculture du département de l'Oise, de celle d'Agriculture de Lyon, Associé libre de la Société des Pharmaciens de Paris, etc.

TOME PREMIER.

A PARIS,

Chez { DENTU, Libraire, rue du Pont de Lodi, No. 3;
L'AUTEUR, au Muséum d'Histoire Naturelle (Jardin des Plantes).

Lamarck's *Philosophie zoologique* was read all over Europe, but his ideas were largely rejected until their merits were reconsidered after his death.

DARWIN AND WALLACE

Charles Darwin agreed with some of Lamarck's ideas. Like Lamarck, Darwin came to believe that species changed over time. However, he did not think that Lamarck's use and disuse theory explained how species changed.

In 1831, at the age of twenty-two, Darwin was invited to spend five years on a British ship called the HMS *Beagle*. He was asked to go along on the voyage as the ship's naturalist. Part of the objective of the ship's mission was to explore and map the coast of South America. The expedition's journey would include a group of islands called the Galápagos Islands,

This copy of the front image of Charles Darwin's 1839 published journal depicts the HMS *Beagle*.

which were about 600 miles (966 kilometers) off the South American coast. Darwin kept careful notes, including drawings, about the animals and plants he saw on his trip. He also collected many fossils and specimens that he brought back to England to study. He was struck by the similarities and differences of some of the plants and animals that he found on the Galápagos to those on the mainland of South America.

When he returned to England, Darwin started a systematic study of the specimens he had brought home. In 1839, he published *Journal of Researches into the Natural History and Geology of the Countries Visited During the Voyage of H.M.S.* Beagle *Round the World*. This book, which is commonly known as *The Voyage of the* Beagle, described his general observations about the trip. However, he said nothing about his ideas of the evolution of species yet. In fact, Darwin kept his theory of how species change to himself for another twenty years. He was aware of how his controversial ideas would upset many people.

The year after Darwin returned from his voyage aboard the *Beagle*, he read the work of British economist Thomas Malthus (1766–1834). Malthus proposed that the human population would grow out of control if there were no natural disasters, disease, and famine. This idea caused Darwin to wonder if the same notion was true of animal populations. Before this time, Darwin had assumed that animals produced only enough offspring to keep the number of animals in the population steady. After reading Malthus's views, however, he began to see that animal populations often produced more offspring than was needed. These offspring had to compete for limited food and resources and not all survived. Quickly, Darwin realized that some of the offspring produced would be better suited to survive their environment than others. Those that were better suited to survive would grow up, mature, and reproduce. In time, the traits carried by these survivors would become more common in the population. Darwin's theory of how species change over time is known as natural selection.

However, even this concept was not Darwin's alone. In 1848, another British naturalist, Alfred Russel Wallace (1823–1913), visited South America. As Wallace traveled through Brazil and the Amazon Basin, he collected samples of plants and animals. Although Wallace's samples were destroyed in a fire on the way back to England, he and his observations survived.

In 1854, Wallace once again set out on an expedition. This time, he traveled to the Malay Archipelago (now Malaysia and Indonesia), where he stayed for eight years. While he was there, he compared the islands' plant and animal species to those of nearby Australia. The similarities between the species led him to believe that they were related, but their differences convinced him that they had evolved differently. Moreover, like Darwin, Wallace discovered and read Malthus's work. Based on his observations and reading, Wallace developed the theory of natural selection independently of Darwin. In July 1858, both Wallace and Darwin published papers describing their theories. And in 1859, Darwin's most famous work was published, fully titled *On the Origin of Species by Means of Natural Selection, or the Preservation of Favoured Races in the Struggle for Life.*

COURTING CONTROVERSY

Charles Darwin began to work on his theory of evolution by natural selection in the early 1840s and did not produce his book about it until the late 1850s. Some attribute his slow pace to hesitancy to reveal his ideas. Why? He realized that many people would consider his theory a challenge to their beliefs about divine creation. But in 1858, Alfred Russel Wallace wrote to Darwin with his own ideas about species evolving. The men decided that summaries of their work would be presented together to the Linnean Society of London in July 1858 so that they could both be credited for their research. From there, Darwin went on to publish his work as the book *On the Origin of Species by Means of Natural Selection* and waited for the reaction in a far-flung part of England. He said the months of anticipation before publication were like "living in Hell."

The book did create shockwaves. For many, the Darwinian view of life—a life of struggle and constant change—conflicted with the creation story of the Bible and implied that life develops through natural processes unrelated to the intervention of God. It challenged the idea that God had created all species of plant and animal life exactly as they are. However, Darwin's seemingly radical theory was actually accepted rather quickly among the scientific community. And many religious communities, too, embraced it, seeing God's hand at work even through natural selection.

Since Darwin's time, a wealth of evidence in the form of fossils, genetics, and discoveries in other fields of science have come largely to support Darwin's main contentions. Yet, some still question evolution today and do not want it taught in schools—or suggest that it only be taught alongside ideas such as intelligent design or creationism.

Even as many people began to accept evolutionary ideas, others mocked Charles Darwin (shown here with an ape body) and his theory.

ADAPTATION

In 1859, Darwin knew nothing of DNA and genes. Gregor Mendel (1822–1884), the "father of genetics," would not publish his research of inheritance "factors" in pea plants for another seven years. It would be decades later when scientists discovered that Mendel's factor was the DNA molecule. However, Darwin did notice something interesting about some of the birds on the Galápagos Islands. In most ways, the finches on the islands looked similar to one another. However, the size and shape of their beaks were different. He noticed that finches that ate seeds on the ground tended to have large, wide, and flat beaks that were good at crushing seeds. The finches that fed on the prickly pear cacti had long, pointed ones. From their similarities, Darwin theorized that the birds were descendants of a common ancestor that was most likely from the mainland of South America.

As the birds spread through the islands, they adapted to different environments on different islands. An adaptation is a change that helps an organism become better suited to its environment. The cactus finch, for example, feeds on the flowers, fruit, and insects associated with the prickly pear cactus. The shape and size of their beaks helped the cactus finches benefit from a food source that other finches could not reach. From their original common ancestor, the Galápagos Island finches would eventually evolve into fourteen separate species.

In 2004, scientists at Harvard University linked Darwin's theory to modern genetics research when they identified a protein molecule that controls the gene that determines the shape of the Galápagos finches' beaks. The molecule, called calmodulin, gives a signal to turn a gene off or on, therefore regulating beak length. Specifically, calmodulin was found at higher levels in cactus finches compared to ground finches, which the researchers said meant elongated beaks in cactus finches. The scientists plan to continue to study calmodulin's role in the development of other Galápagos finch beaks as well as the protein's role in human genetics.

You can see the differences in the beaks of these four birds that became known as "Darwin's finches."

CHAPTER 3

EVOLVING POPULATIONS

Following Darwin's natural selection theory, other biological disciplines arose that bolstered or added to his ideas. This blending of scientific work from evolutionary theory to genetics is called modern synthesis. The term comes from zoologist Julian S. Huxley's 1943 book *Evolution: The Modern Synthesis*, but over the years, many other scientists have contributed to the combination of the ideas that inform our understanding of evolutionary biology.

HOW A SPECIES CHANGES

One of the scientists who helped advance modern synthesis was Ernst Mayr (1904–2005). In his book *Systematics and the Origin of Species from the Viewpoint of a Zoologist* (1942), Mayr proposed that a group of organisms separated from its main group by either time or distance could adapt in such a way as to eventually make the two groups incapable of interbreeding. If this occurs, a new species can be formed. If this process occurs because of a physical separation, it is called geographic speciation.

Ernst Mayr was one of the twentieth century's foremost evolutionary scientists. His work contributed to modern ideas of genetics and Darwinian evolution.

Consider, for example, that a group of coastal rabbits are living and breeding together when a huge storm comes along and washes some rabbits out to sea. Luckily, the ocean-going rabbits are able to float along on trees that were swept out to sea. Eventually, the marine rabbits wash up on a deserted island. Now there are two groups of rabbits. These two groups can no longer mate with one another. The gene flow, or exchange of genetic information, between the two groups is cut off.

The environments that the two groups now live in are different. The island rabbits adapt to their new environment. Maybe some of the rabbits adapt better than others. Over time, natural selection works on this group of rabbits. The ones that have best adapted to the food and climatic conditions on the island survive and reproduce. The others die out.

At the molecular level, the two groups are also evolving differently. Varying genetic mutations accumulate in the two distinct populations. For example, let's say that a mutation appears and spreads among the island rabbits. This mutation changes the rabbits' eating habits. The change in eating habits makes it easier for the island rabbits to survive on the island's plants, which are different from the mainland plants. The rabbits with this mutation have an advantage over the rabbits that do not have the mutation. They eat more food and the nonmutated rabbits start to die out. Over time, more and more baby rabbits are born with the mutation.

If, after several centuries of this isolation, another storm blows through and washes some of the island rabbits back to shore, they may

encounter the mainland rabbits. However, if in their time on the island, the island rabbits have become too genetically different from the mainland rabbits, the two populations may not be able to interbreed. It is possible that the island rabbits have evolved new mating rituals. Or they may reproduce during a different season. Or maybe the two groups of rabbits try to mate, but produce no young. If any of these scenarios proves to be true, the two groups of rabbits have evolved into two separate species.

Even if two groups of rabbits look similar, they are considered different species if they cannot produce young with each other.

AN ICONIC EXAMPLE OF EVOLUTION

The peppered moth provides a strong example of how natural selection works. Before the Industrial Revolution occurred in Great Britain in the 1700s, many of the region's trees had light-colored bark. The peppered moth, which spends daylight hours resting on tree trunks, was a light, creamy color. This made the moth challenging to see against the tree bark, which provided it protection against its predators. However, the Industrial Revolution brought coal-burning factories to Britain, sometimes covering nearby trees with soot. This soot killed lichens and mosses on the tree bark, which made the trees look darker. The light-colored moths, called *Biston betularia morpha typica*, could be easily seen against the dark bark now. Around 1848, people noticed a dark-colored variety of moth. Around fifty years later, about 98 percent of peppered moths seen in Manchester were these dark-colored moths, called *Biston betularia morpha carbonaria*. Because the *carbonaria* type was harder to see than the *typica*, they were eaten less often than the *typica*. While *typica* did not become extinct, it was rarely seen except in rural areas, where trees were less affected by the pollution of industries. However, when clean air laws were passed in the twentieth century, reduced air pollution meant the return of lighter colored trees—and *typica*. It is now uncommon to see *carbonaria* in Britain.

In 2016, researchers from the University of Liverpool announced in the journal *Nature* that they had pinpointed the genetic mutation that allowed the dark color in the moths to emerge. They even calculated a likely date for the mutation: 1819. It was not an easy feat, however. They studied black and pale moths in an effort to find the genetic markers that would have been responsible for the dark coloring. They focused on 400,000 bases in a certain part of the moths' DNA. Among these, they identified all the differences they could find between the two colors of moths, a process one of the scientists called

"excruciatingly tedious." When they eventually found the mutation, they discovered it was an unusual kind that often can be damaging to the gene. However, in this case, it helped the moths after it was introduced to the population. It should be noted that, while the research team found the mutation, how it causes the dark coloring is still unknown. And in fact, the gene they found it in, called cortex, has not been known in the past to have anything to do with pigment.

The process of organisms changing color to blend in with industrialized environments is called industrial melanism, and it is observed in different types of moths, butterflies, and beetles. British scientist H. B. D. Kettlewell remarked in 2009, "[If] Darwin observed industrial melanism he would have seen evolution occurring not in thousands of years but in thousands of days."

A mutation followed by natural selection resulted in a population of moths with darker coloration in a short period of time.

GENE FLOW, GENETIC DRIFT, AND THE FOUNDER EFFECT

Natural selection is not the only way evolution occurs. Sometimes genetic variation is introduced into a population through migration. For example, consider two populations of lizards. One population carries the allele for green skin color. The other lizard population has brown skin. If one or more of the brown lizards moves to the area where the green lizards live and reproduce, the allele for brown skin color has been introduced into the population. The introduction of genes into a population due to the movement of an individual from one population to another is called gene flow.

The frequency of an allele in a population may also change if individuals with a certain trait and the alleles that carry that trait die or fail to reproduce, just by chance. Then only the survivors can pass their genes on to the next generation. If the population contained few individuals to begin with, the accidental loss of those individuals causes a measurable change in the frequency of the allele. Such random changes over time are called genetic drift. Genetic drift can cause evolution and speciation in a population. However, unlike natural selection, genetic drift does not select for better-adapted or healthier individuals. The process is entirely random and is noticeable only in small populations.

For example, consider a tiny population of lizards that contains both brown and green individuals. At first, assume that there are equal numbers of brown and green lizards.

Then, during one generation, several of the green lizards are crushed by a falling boulder. Now the allele that controls green skin coloration is reduced in the population. Just by chance, the ratio of brown genes to green genes has increased. More individuals carrying the brown skin coloration have survived to have offspring. The change in frequencies of the alleles constitutes, by definition, evolution. Genetic drift reduces the genetic variation in a population.

Unlike the lizards in the example, chameleons have evolved to be able to change their coloring.

The smaller the population, the faster the effects of genetic drift can be felt. If in future generations, for example, some other natural disaster should befall the small population of green lizards, the gene that produces green skin could be lost forever.

Another example of genetic drift is the so-called founder effect. The founder effect occurs when a small population, the "founders," starts a new colony in a different place, with the founders having one or more differences in gene frequency from the larger population as a whole. Because the founder population is small, distinctive alleles that they carry can, over time, be lost entirely through genetic drift—or they may increase in frequency, also by genetic drift.

The Old Order Amish community of eastern Pennsylvania provides an example of the founder effect. These people are the descendants of about two hundred German immigrants that settled in the area in the 1700s. In this population of immigrants, some genes had a higher frequency than others. Because the Amish do not marry outside of their small population, the frequency of some genes has gotten even higher over time. For example, the recessive genetic condition often called Ellis-van Creveld syndrome is much more common in the Amish community than it is in the general American population. This genetic condition causes polydactyly (extra fingers and toes), dwarfism, and heart abnormalities. Scientists have been able to trace the recessive allele that causes Ellis-van Creveld syndrome back to one couple that came to North America in 1744. The allele was passed down through generations of their descendants. As the offspring of this couple married and reproduced, the allele increased in frequency. About seven in one million people are born with Ellis-van Creveld syndrome to people in the general population of the United States today. However, in the Old Order Amish community, five in one thousand babies are affected by the condition.

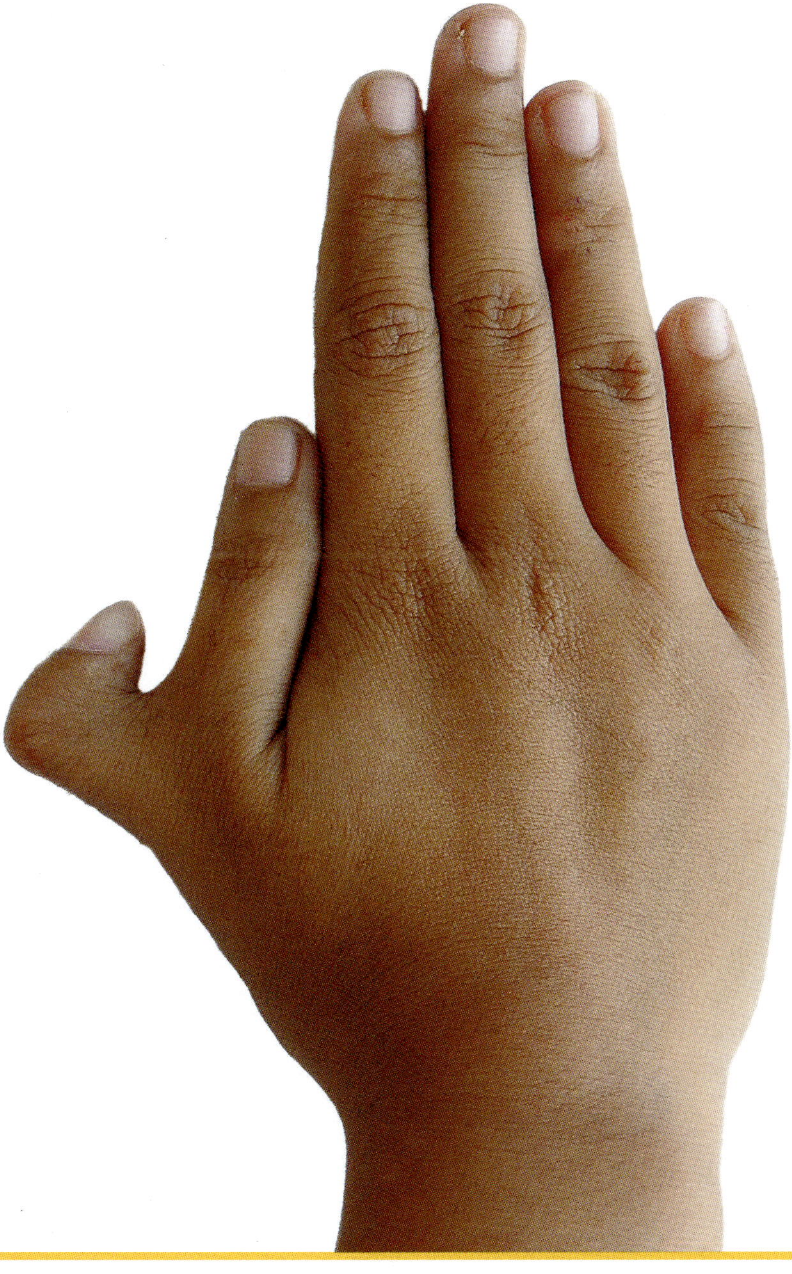

Occurrences of polydactyly are more common in the American Amish community than the greater U.S. population because of the founder effect.

THE HARDY-WEINBERG PRINCIPLE

Natural selection, gene flow, genetic drift, the founder effect, and mutation are all ways that the frequency of an allele can increase or decrease in a population. In the early twentieth century, two scientists, Godfrey Hardy and Wilhelm Weinberg, proposed a mathematical model that explained how genotype frequencies change from generation to generation. This mathematical model has become known as the Hardy-Weinberg principle. The Hardy-Weinberg principle states that in a large population with random mating, allele frequencies remain constant in the absence of disrupting factors that could cause the ratio to change. This condition is known as genetic equilibrium.

For genetic equilibrium to exist, several conditions must be met, including the following:

- The population size must be very large.
- Mating must be random.
- All members of the population breed.
- All matings produce an equal number of offspring.
- Members of the population cannot leave and outsiders cannot get in.
- There are no mutations.
- There is no natural selection.

If just one of the seven conditions of the Hardy-Weinberg principle is not met, the frequencies of alleles in the population will change, and evolution, by definition, occurs. It is extremely uncommon that all seven of these conditions are met in a large population, though. For example, in what conditions would individuals in a large population not be able to leave and outside individuals not be able to get in? Certainly there are possibilities, such as on remote islands, but still, genetic equilibrium remains rare. So, most populations are evolving constantly.

Most populations in nature violate at least one of the assumptions of the Hardy-Weinberg principle and so are seldom in genetic equilibrium.

CHAPTER 4

EVIDENCE FOR EVOLUTION

For people who are not scientists, the word "theory" is used as a synonym for "guess." But for scientists, a theory is almost the opposite. A theory without the support—the evidence—is just an idea. Theories can evolve, however, as scientists gather evidence to prove—or disprove—them. As new evidence is collected, they revise and refine their conclusions.

Evidence collection is vital to the study of evolution. To reconstruct the history of living organisms, scientists have gathered evidence and used it to determine how species might be related to one another. Over time, their revisions of hypotheses have aided scientists in creating a better understanding of the way organisms are connected. Studying physical traits, behavioral traits, or genetic sequences have helped them describe the branches of the evolutionary tree.

THE FOSSIL RECORD

The fossil record is one type of data scientists rely on to tell them the story of organisms that lived in the past. Fossils may be made up of an organism's actual remains, such as bones, teeth, or shells. Or they may be the remains of some characteristic of an organism's behavior, things like nests, footprints,

or burrows, for example. By studying fossils, scientists can learn many details about the animals and plants that lived on Earth thousands or even millions of years ago.

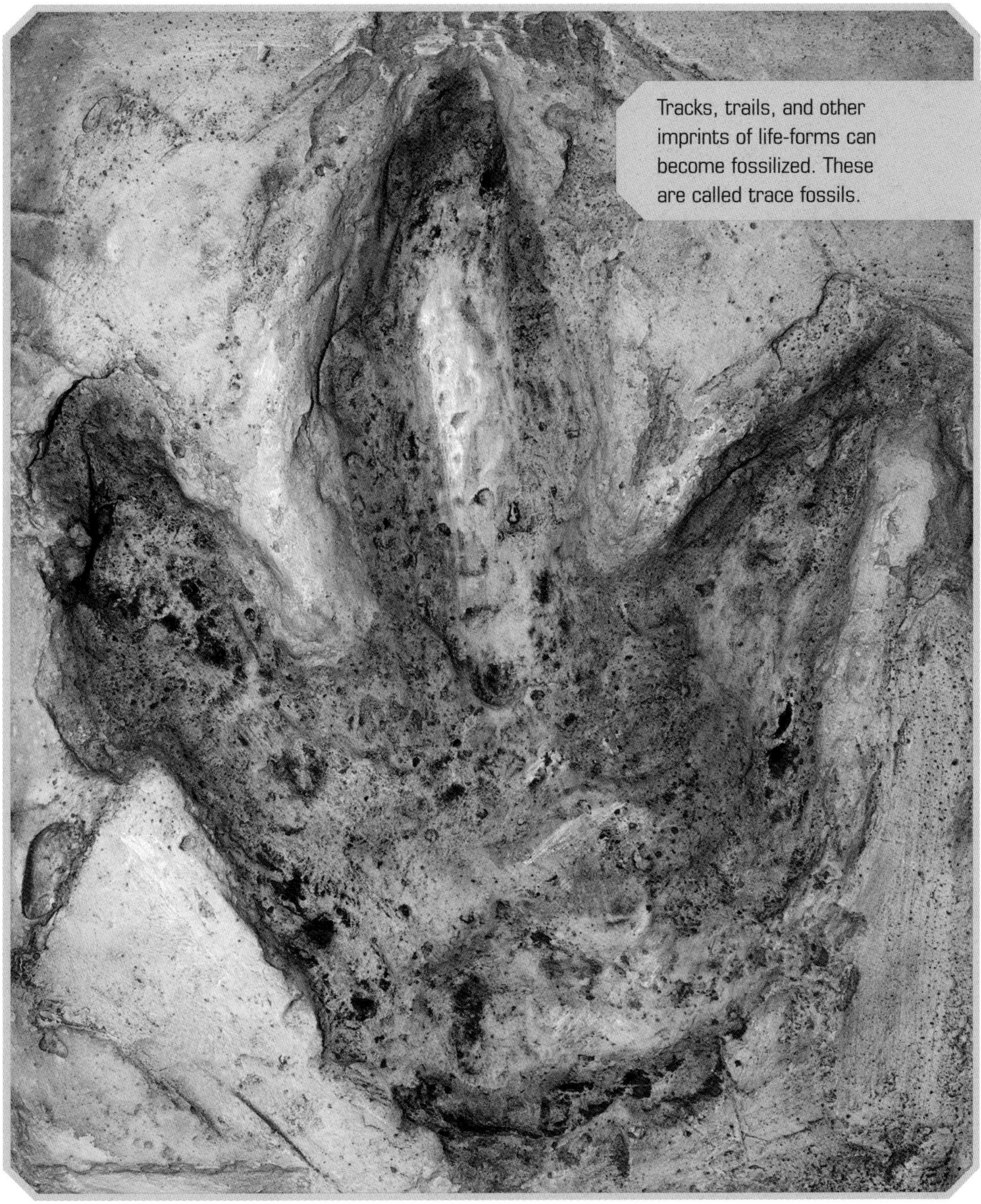

Tracks, trails, and other imprints of life-forms can become fossilized. These are called trace fossils.

For a fossil to form, certain conditions must be present. And even if these conditions do exist, only parts of a plant or animal will likely survive to tell its story to generations of people millions of years later. Hard parts of the body, such as bones, teeth, or shells, are the most common parts that are fossilized. It is much more difficult for scientists to find soft tissue, such as skin or organs.

About three-fourths of Earth's land surface is covered by sedimentary rocks. Sand, mud, or other sediments form sedimentary rock. Over time, layer after layer of sediment settles on top of one another. The layers are pressed further and further into Earth until, eventually, the bottom layers turn to rock. Sometimes a dead animal or plant gets caught in one of the layers and forms a fossil.

Occasionally, the conditions are just right, and some soft parts of an animal's body are preserved. In rare cases, even the whole body of an ancient plant or animal is found. For example, in 2020, Russian scientists found a preserved baby woolly rhinoceros in the frozen tundra of Asia. They estimate that the extinct Ice Age animal lived between twenty thousand and fifty thousand years ago.

Woolly rhinoceroses were covered with long hair, an adaptation that allowed them to survive in their harsh Arctic habitat.

WHAT ARE LIVING FOSSILS?

The term "living fossil" is meant to describe an organism that has remained somewhat unchanged over millions of years or an organism that has no, or very few, close surviving relatives. Cockroaches, sharks, horseshoe crabs, and tuatara are examples of animal living fossils. Ferns and cork palms are examples of plant living fossils. However, it must be stressed that every species existing today has evolved since it first appeared; it just may not have evolved to a great degree. Charles Darwin was the first to use the phrase "living fossil." He meant that such a species did not need to adapt and evolve much because it was living in an environment that suited its survival.

The oldest fossil of a horseshoe crab, for example, dates back to more than 400 million years ago. Its features are basically the same as back then. The horseshoe crab has no close relatives. In fact, it is more connected to spiders in its distant past than it is to crabs.

The tuatara looks like a lizard but is very different. The tuatara's closest relatives are an extinct group of reptiles. The tuatara is the last living species of the group known as the Rhynchocephalia, which evolved sometime in the early Triassic period, when dinosaurs first began to appear.

The cockroach, the oldest living winged insect, has been around for about 350 million years. Like the horseshoe crab and the tuatara, cockroaches roamed Earth at the same time the dinosaurs did. Their fossils are nearly the same as the creatures we see today.

Scientists once called sharks living fossils, but fossil evidence is proving that the label is not accurate. The study of a 325-million-year-old shark, which had a gill structure more like bony fishes, suggests that modern sharks have actually evolved quite a bit over the years.

Some who do not believe in evolution use the animals labeled by scientists as living fossils as so-called proof that evolution does not exist for any species. So why did all these species survive while the dinosaurs did not? Because they were able to adapt to changes in their environment faster and better than the dinosaurs were.

Horseshoe crabs (fossil shown above) have sharp tails, but they are harmless. They use their tail to flip themselves back over if they are overturned.They have been doing this for millions of years!

Early naturalists did not want to believe that plants and animals could become extinct. Therefore, when they found a fossil, they would try to make it match up with a living species. Even if they compared the structures in the fossil to that of a living organism and found differences, they would convince themselves that the fossil was just an unusual or misshaped specimen. Near the end of the eighteenth century, however, French scientist Georges Cuvier (1769–1832) showed that separate rock layers contained distinct types of fossils. He also showed that the deeper the rock layer, the more different the fossils were from living species. In general, deeper rock layers correspond to earlier time periods than those near the surface. Cuvier was skilled at piecing together the animal skeletons he unearthed. He was also very good at showing how those skeletons were different from the animals that lived at the time. Using these differences, Cuvier was able to convince skeptics that some plants and animals that lived on Earth had, indeed, disappeared forever.

Using fossils, scientists can attempt to piece together how life on Earth has evolved. They have, for example, traced the modern horse back to a four-toed animal that lived about 55 million years ago, applying fossil evidence. Fossils have shown that the earliest horse ancestor was about the size of a fox terrier. Scientists call this animal *Hyracotherium*. However, it is better known by the name *Eohippus*, which means "dawn horse." *Eohippus* fossils have been found in the western United States and in Europe. The large tree of horse evolution contains at least fifteen branches. Most of those branches lead to dead ends. In fact, *Equus*, the genus of the modern horse family, is the only surviving one. This branch of the tree developed about 4 million years ago, and it is still going strong. The modern horse family contains seven species, including the domestic horse, zebras, donkeys, and other animals.

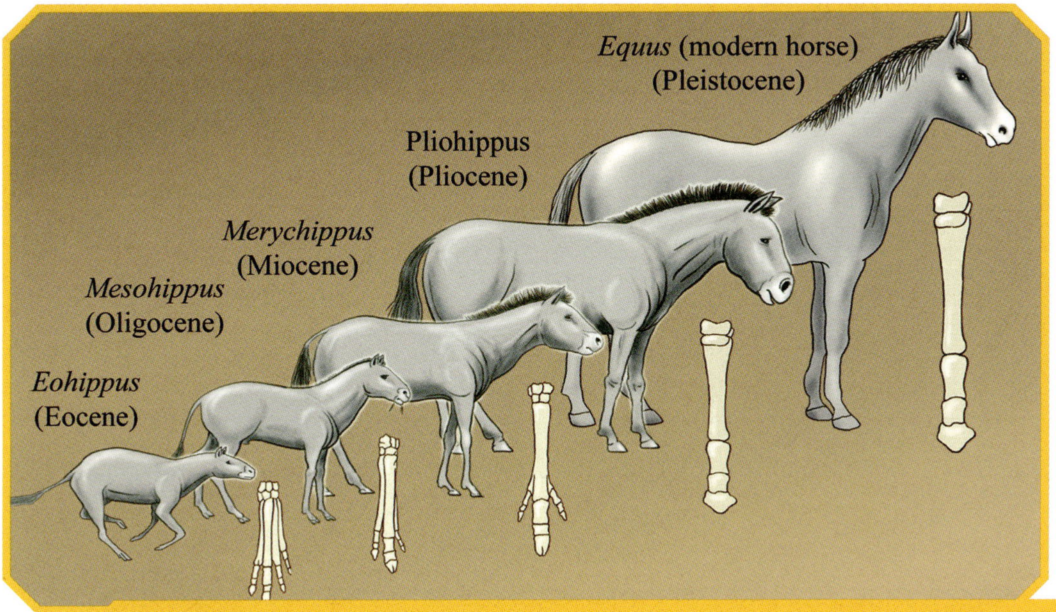

Equus disappeared from North America about 8,000 years ago for an unknown reason. It did not appear again until Spanish explorers brought horses to the continent in the 1500s.

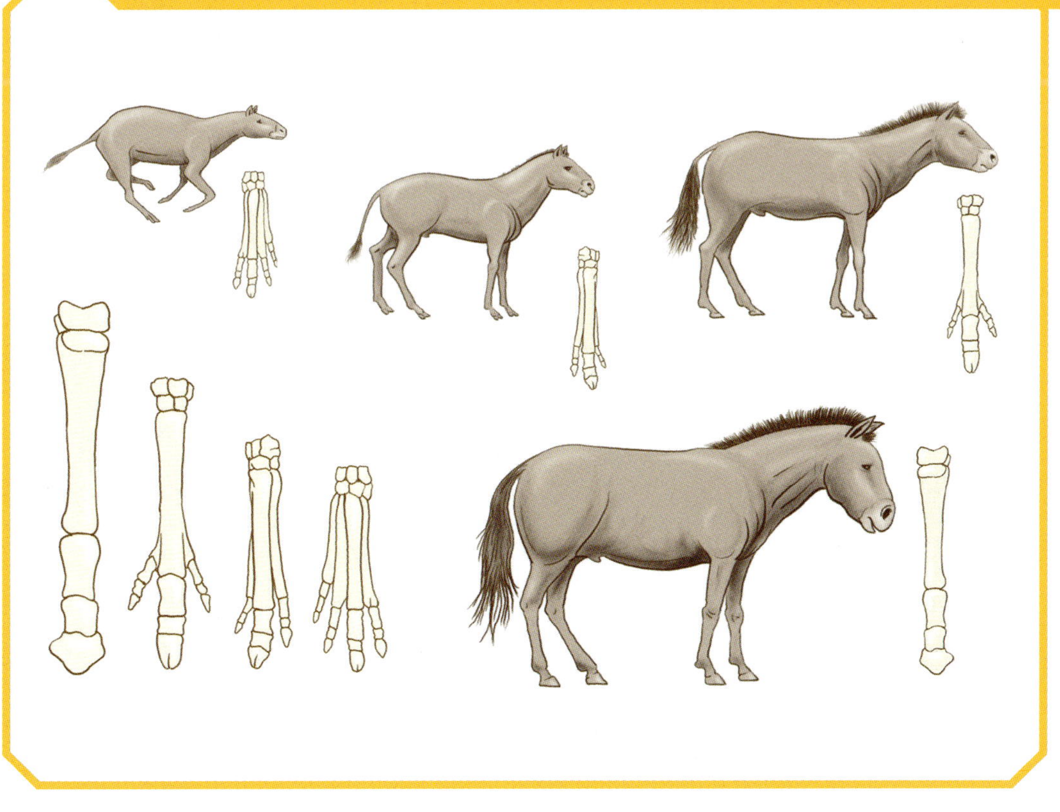

TRANSITIONAL FOSSILS

One criticism often used to discredit the use of the fossil record as evidence of evolution is the so-called missing link. A missing link is a fossil that shows characteristics of two different species; it could be considered the link between them. A fossil that shows the characteristics of a human and an ape, for example, would be a missing link. Scientists, however, prefer to use the term "transitional fossil" because the term "missing" implies that a fossil should exist but that it does not. Fossilization is an iffy process. Most organisms that die do not do so under the conditions necessary for fossil formation. Therefore, scientists may never see these organisms in the fossil record. That does not mean that they did not exist. It just means that they were not fossilized. However, sometimes scientists get lucky and a transitional fossil is found. These transitional fossils can help scientists discover the evolutionary steps required for one species to evolve into a new one.

For example, scientists had long hypothesized that the first tetrapods, animals with a backbone and four limbs, lived in the water. They had evidence that the first land-dwelling tetrapod crawled out of the water about 375 million years ago. Scientists began to suspect that land-dwelling tetrapods may have evolved from animals called elpistostegids. These animals were large meat-eating fish that are now extinct. Researchers believed that these fish would have been a lot like today's alligators or possibly large salamanders. For years, scientists looked for fossil elpistostegids to study. They did not find many and the few that they could find were not complete.

Then, in 2004, researchers Neil Shubin, Edward Daeschler, and Farish A. Jenkins found a well-preserved elpistostegid fossil on Ellesmere Island in northern Canada. They named the fossil *Tiktaalik roseae*. *Tiktaalik* had scales and gills like a fish. However, it also had thick ribs and strong neck, shoulder, and wrist bones that could be used to prop itself up on land. Scientists hypothesize that these bones developed when *Tiktaalik* adapted to prey on fish in shallow water. In the process, the animal took on some characteristics needed to transition from a swimming fish to a land-dwelling creature.

As a transitional fossil, *Tiktaalik* gives scientists a peek at the stepping-stones of evolution.

Tiktaalik roseae is not the only transitional fossil that scientists have found. In fact, one of the first, *Archaeopteryx*, was described about two years after Darwin's *On the Origin of Species*. In 1861, an *Archaeopteryx* fossil was found in 150-million-year-old rocks. These rocks were formed during the late Jurassic period, when the dinosaurs lived. The fossil showed an animal with the teeth and long bony tail of a reptile, but with the feathers of a bird. Many other fossils of feathered dinosaurs have been found since then too.

Scientists think the largest species of *Archaeopteryx* could grow to about 1 foot 8 inches (0.5 m) in length.

In 2009, scientists unveiled what they thought could be another transitional fossil. This time the fossil was found in Germany's Messel Pit, a site that has yielded many well-preserved fossils. Scientists call the small fossilized female primate *Darwinius masillae*. She has been nicknamed Ida. Ida lived about 47 million years ago. Originally, some scientists believed that Ida may have been one of the transitional fossils between simians (monkeys, apes, and humans) and a more distant lemurlike relative. Ida had hands that could have grasped objects, fingers that had nails instead of claws, and opposable thumbs (thumbs that can bend to touch the other fingers) like both simians and lemurs. But the rest of her skeleton is more like that of a lemur. Further studies have shown that Ida was probably a dead-end branch on the lemur part of the primate family tree rather than a close simian relative.

HOMOLOGY AND SYNAPOMORPHY

Using fossils is not the only way scientists compare characteristics of related organisms. The word "homology" was first used in the 1800s by British scientist Richard Owen. Owen observed similarities between some structures in organisms, such as the likenesses of limbs. Later, Charles Darwin suggested that such resemblances could have arisen from a common ancestry between the organisms.

Because the term was coined in a pre-evolutionary era, its meaning is sometimes confused and confusing. Today, most scientists prefer the term "synapomorphy" to "homology." A synapomorphy is a characteristic shared by all descendants of a common ancestor, but not shared with other groups. ("Synapomorphy" comes from Greek words meaning "shared shape or form.") Such features can be discovered by comparing the anatomies of different species, examining similarities and differences at the cellular level, or studying vestigial structures.

By studying anatomy, for example, scientists have discovered that whales, bats, and humans all evolved from a common ancestor. The forearms of all these animals may look vastly different on the outside, but they look remarkably similar under the skin. Structures, such as a whale's flipper, a bat's wing, and a human's arm, that have similar anatomy but different functions are called homologous structures. The prefix *homo* means "same." Organisms with homologous structures most likely shared a common ancestor somewhere in their evolutionary pasts.

Sometimes animals have structures from a common ancestor that they no longer use. Notice the hind legs (spurs) on this boa constrictor.

Another piece of evidence that has convinced scientists that all living organisms have evolved from one common ancestor is that, essentially, they are all alike. For example, all living beings are made up of cells. These cells have a cell membrane that contains a mix of molecules including lipids (fatlike substances that do not dissolve in water), proteins, and carbohydrates. Of these molecules, 99 percent are made up of only six elements. In addition, all plants and animals contain DNA. This genetic molecule is remarkably similar in all living things. No matter whether the DNA molecule is in a bacterium, a rat, or a human, it is still made up of a sugar-phosphate backbone and the four bases: adenine, guanine, cytosine, and thymine.

Researchers have realized how some structures inside cells have evolved. Some of the earliest cells on Earth were single-celled bacteria. Certain types of those primitive bacteria made their way inside more complex organisms. Indeed, they still live inside every living multicellular organism today. Scientists believe that a cell structure in animals (and plants and fungi and other multicellular organisms) called a mitochondrion and a structure in plants called a chloroplast started out as independent bacteria. They believe this notion because these are the only structures in the cell that have their own DNA just as bacteria do. They also reproduce, copying their own DNA, just as bacteria do. Their DNA can be traced back to the bacteria that were their ancestors. At some point, these smaller cells came to live inside larger ones. This idea is called the endosymbiotic theory of evolution. Two species have a symbiotic relationship when they live and work together for the benefit of both species. And when one organism lives inside another organism and functions as a single organism, it is called endosymbiosis.

CHAPTER 5

THE EVOLUTION OF HUMANS

Some people do not understand how evolution works. They think that people evolved from gorillas because humans and gorillas are said to be on the same evolutionary branch. This is not true. Evolutionary scientists would not support this theory. Instead, they say that humans and gorillas share a common ancestor, an ancestor that was neither human nor gorilla.

Scientists also make no claims that humans are more evolved than gorillas (or other primates). Instead, primate species have evolved in different ways, unique to each species.

EVOLUTION AND PEOPLE

Modern humans have been around for about the last one hundred thousand years or so. Scientists call the group of bipedal primates—which include chimpanzees, gorillas, and humans—hominids. The ones that belong to the human lineage are called hominins. Over millions of years, the hominin line branched into at least twelve different species. However, only one remains: *Homo sapiens*, or modern humans.

The first hominin fossil to be discovered was found in 1856, three years before the publication of Darwin's *On the Origin of Species*. The skeleton was found in a cave in the Neander Valley of Germany. This early group of humans has become known as *Homo neanderthalensis*, or the Neanderthals. They lived about 230,000 to 30,000 years ago, and their time on Earth overlapped with groups of *Homo sapiens* with more modern human features.

This is the skeleton of a four-year-old Neanderthal child found buried in France, originating from 75,000 years.

Thirty-seven years after the first hominin fossil was found, the first example of *Homo erectus* was discovered in Indonesia on the island of Java. Scientists nicknamed this fossil Java Man. *Homo erectus* is believed to have lived 1.8 million to 200,000 years ago. Since the discovery of Java Man in 1893, many other examples of *Homo erectus* fossils have been found throughout Africa and Asia. Scientists believe that *Homo erectus* is likely the first hominid to have used fire.

In 1924, Australian anthropologist Raymond Dart discovered an apelike skull with humanlike teeth in South Africa. This ancient human relative was given the name *Australopithecus africanus*. It had a larger brain than chimpanzees, but a

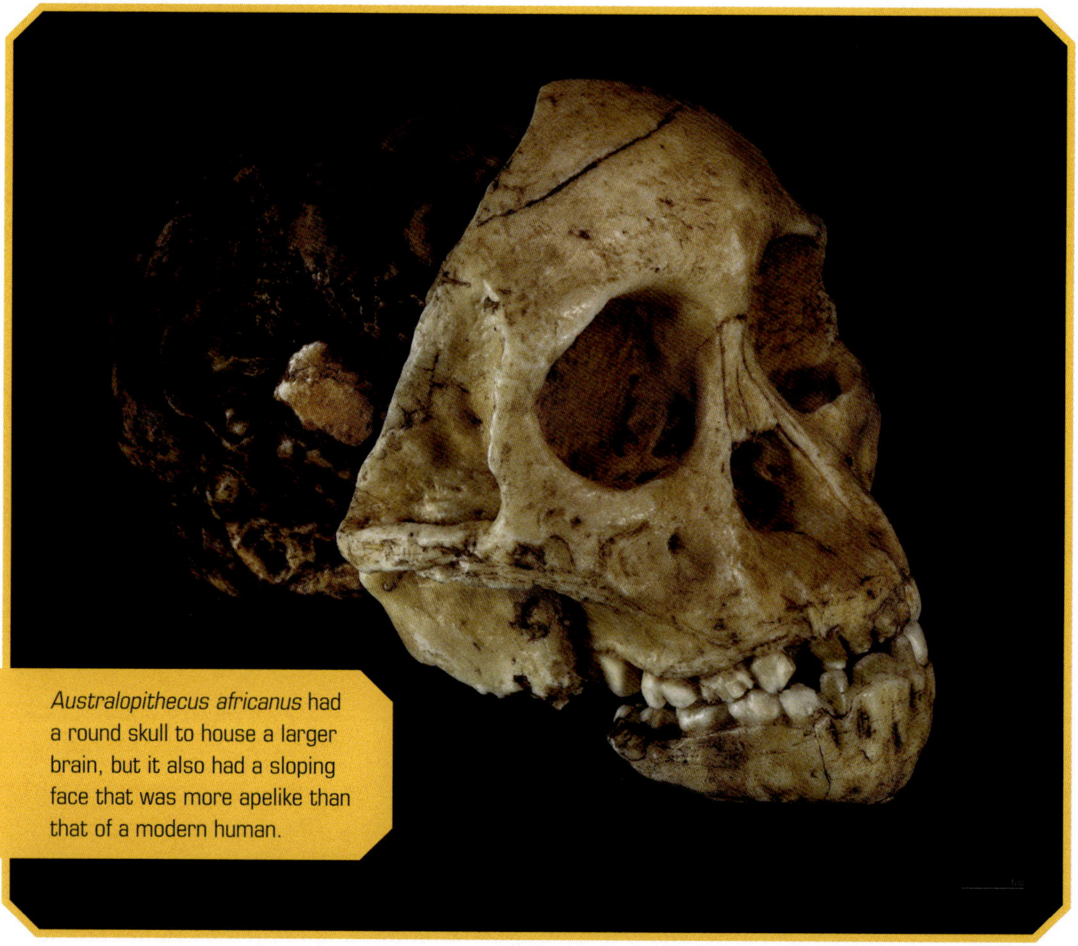

Australopithecus africanus had a round skull to house a larger brain, but it also had a sloping face that was more apelike than that of a modern human.

smaller brain than modern humans. However, the specimen that Dart found was the skull of a child who was about six years old. Because of the similarities of a human child's skull to that of a young ape, his find was largely overlooked. Dart's discovery was nicknamed the Taung Child.

Another reason Dart's Taung Child was basically ignored was that it was very different from the other fossils found. One of those fossils was called the Piltdown Man. Piltdown Man was unearthed in the Piltdown quarry in Sussex, England, by amateur geologist Charles Dawson. The fossil appeared to have a humanlike skull and an apelike jaw. Dawson showed the bone fragments to Arthur Smith Woodward, the head of the British Museum's geology department. Woodward announced to the Geological Society of London on December 18, 1912, that the "missing link" had been found. Almost immediately, some scientists were skeptical. They argued that the skull and jaw were not from the same individual, but were actually separate human and ape fossils. But then in 1913, Woodward and Dawson found another fossil. This time the fossil had a canine tooth that was in between the size of an ape's canine and a human's. In 1915, the pair announced they had found more remains several miles from the original Piltdown site. The fossil molar and skull fragments appeared to be similar to those of Piltdown Man. This discovery was named Piltdown Man II.

However, in the 1920s and 1930s, many more hominin fossils were discovered, including the Taung Child. These had little in common with the Piltdown fossils. Indeed, as more fossils were found, it became clear that the Piltdown fossils did not fit into the progression of hominin evolution. Scientists were puzzled. At least they were until the late 1940s when chemical tests were developed that allowed the Piltdown Man fossils to be reexamined. What scientists discovered after these chemical tests was that the bones were not as ancient as they appeared to be. Instead, the skull had been boiled and stained to make it appear as if it were old. And the teeth on the jawbone had been filed down to make them look humanlike. In 1982, scientists were able to determine that the jawbone belonged to an orangutan. The whole thing was a hoax. Exactly who decided to carry out the deception is not known. But once the Piltdown fossils were exposed as a fraud, the rest of the fossils found in the intervening years made more sense.

In the early 1960s, the team led by paleoanthropologist Louis Leakey found the first fossil of *Homo habilis*. *Homo habilis* means "handy man." It was given this name because scientists believe it was the first hominid to use tools. The brain of *Homo habilis* was larger than the brain of any of the examples of *Australopithecus* but smaller than the other *Homo* species. *Homo habilis* lived 2.4 to 1.5 million years ago.

THE LEAKEY FAMILY

Paleoanthropologist and archaeologist Louis Leakey was born on August 7, 1903, in Kabete, Kenya. His parents were British missionaries who raised Leakey among the Kikuyu people of Kenya. Though a university professor had advised Leakey to look for early humans in Asia, in 1924 he began his work in East Africa. Leakey wanted to believe that Charles Darwin was correct when he suggested in his 1871 book *Descent of Man* that early humans likely lived in Africa because their nearest evolutionary relatives, chimpanzees and gorillas, lived there. Leakey was joined in his work by his wife, the anthropologist Mary (Nicol) Leakey, and their sons. In fact, Mary made many of the most famous Leakey discoveries herself.

In 1959, Mary uncovered a hominin fossil that was given the name *Zinjanthropus* (considered now to be a form of *Paranthropus*, similar to *Australopithecus*). It was believed to be about

1.7 million years old. And it was Louis and Mary's son Jonathan Leakey who first discovered *Homo habilis* at Olduvai Gorge in Tanzania in 1960. His father believed *Homo habilis* lived at the same time as *Australopithecus* in East Africa and was a more advanced hominin. At first, many scientists found Leakey's interpretations dubious, although they thought the finds themselves were noteworthy. They argued that *Homo habilis* was not different enough from *Australopithecus* to be given a separate classification. Later finds, however, established that *Homo habilis* does represent an evolutionary step between the australopiths (who eventually became extinct) and *Homo erectus*.

The Leakeys went on to make other important finds. Mary and their son Richard and his wife, Meave (and later their daughter Louise), continued the fieldwork while Louis focused on fundraising for other research projects. He founded several prominent organizations, including the Tigoni Primate Research Center in Nairobi as well as the Leakey Foundation to further the study of human origins. Louis Leakey died on October 1, 1972, in London, England.

Louis and Mary Leakey dig at the famous Olduvai Gorge in Tanzania.

In 1974, Donald Johanson found another important *Australopithecus* species. This specimen was given the name *Australopithecus afarensis* and nicknamed Lucy. Lucy was special because she lived about 3.2 million years ago and her anatomy revealed to scientists that humans were walking upright on two feet at that time. In 1975, Johanson made another find: the First Family. This group of fossils was found close to the same place where he had discovered Lucy. The bones belonged to thirteen people, but they were found all together. This was the first evidence that ancient humans lived in groups.

Only 40 percent of Lucy's skeleton was found in Ethiopia, but it was enough to reconstruct what she probably looked like.

In 2009, scientists made yet another big announcement. Between 1992 and 2009, an international team had discovered fossils from a formerly unknown hominid species, called *Ardipithecus ramidus*. This primitive prehuman species lived about 4.4 million years ago, making it even older than Lucy. The most complete skeleton found was that of a female nicknamed Ardi. Like Lucy, Ardi walked upright when she was on the ground. However, her skeleton shows that she had feet that were more suited to climbing trees and a smaller brain than Lucy had. Scientists are still working on how Ardi and her kind fit into the human evolutionary tree.

New discoveries continue every year. In 2020, a South African team led by Andy Herries from La Trobe University reported new fossils of both *Paranthropus robustus* and *Homo erectus* dating to between about 2 and 1.95 million years ago, making them the oldest fossils of these hominin species. These finds demonstrate the two species existed at the same time at this site with *Australopithecus africanus* and also pushes back the origin of *Homo erectus* by about 150,000 to 200,000 years.

YOUR HEALTH AND EVOLUTION

When one species evolves into another through genetic changes over countless generations, the process is called macroevolution. However, sometimes, as was seen in the peppered moth in Britain, evolution does not take that long. When the frequency of a gene changes in a population over a short period of time, the process is called microevolution. Scientists must use evidence, such as fossils, to try to piece together macroevolution. Even so, they can often observe microevolution taking place.

In 1928, Alexander Fleming discovered that bacteria would not grow near the mold *Penicillium notatum*. Fleming correctly reasoned that the mold gave off a substance that killed or slowed bacterial growth. He named this substance penicillin. Fleming had discovered the first antibiotic. Over the next several decades, scientists would find many other substances that would kill or slow the growth of bacteria and fungi. They used these medications to greatly reduce the number of deaths caused by diseases, such as tuberculosis and pneumonia, which are caused by bacterial infections.

However, bacteria have weapons of their own. They fight being wiped out. Some bacteria carry genes that make them resistant to the effects of antibiotics. When people take antibiotics, most of the bacteria invading their bodies die. However, the antibiotic-resistant bacteria survive. Then they reproduce. The frequency of the genes that allow for resistance to antibiotics increases in the bacterial population. Because bacteria can reproduce very quickly, evolution occurs rapidly. As a result, antibiotic-resistant bacteria have become a problem for people. When bacteria become resistant to a certain antibiotic, that drug no longer cures a person's bacterial infection. Some bacteria have become resistant to more than one antibiotic.

In an example of natural selection, bacteria can become resistant to antibiotics such as penicillin over time. The resistant bacteria is shown in blue below. These bacteria survive and reproduce, creating bacteria that are also not affected by the antibiotic. Some resistant bacteria can also give their resistance to other germs (in blue-green).

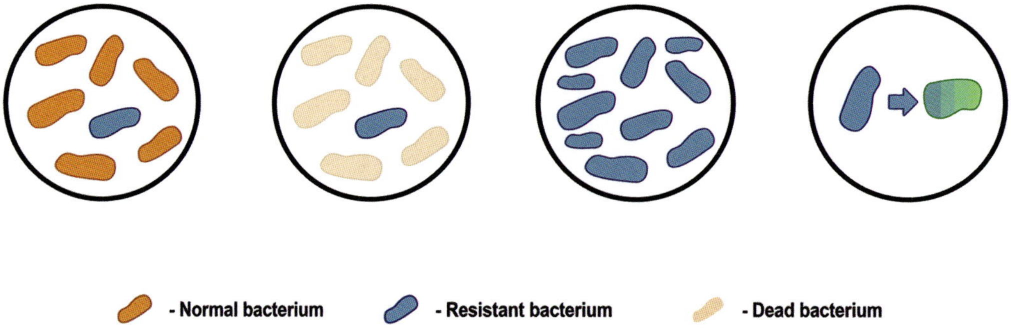

Overusing antibiotics can cause this problem to become worse. Therefore, antibiotics should be taken only when an individual has a bacterial infection. Antibiotics do not work on viruses and will just increase the frequency of antibiotic-resistant genes in the bacterial population. Not using all of the antibiotics prescribed by a doctor can do the same thing. It wipes out the susceptible bacteria and leaves behind the resistant ones to grow more plentiful, possibly making the individual sick again.

THE BENEFITS OF BIODIVERSITY

Genetic diversity within a population is essential for its long-term survival. Without genetic variation, species may not be able to adapt to different ecological conditions. If a species cannot adapt, it will most likely disappear. When species go extinct, their disappearance decreases the variety of life, or the biodiversity, on Earth.

Human actions have a huge impact on biodiversity. When people chop down forests to build houses, for example, they destroy the natural home, or habitat, of many plants and animals. Habitat destruction also occurs when humans claim grassland to create farmland or to build on. When their habitat is disrupted in this way, most animals and plants do not have time to adapt to a new environment. If they cannot adapt, the plant and animal populations decrease. Therefore, there are fewer individuals left in the population in which to mate. Fewer breeding individuals lead to a smaller gene pool and less overall genetic diversity.

Reduced biodiversity can also spell trouble for humans. People get many of the things they need from the plants, animals, and other organisms around them. These organisms provide food, fiber, fuel, and medicines, for example. Scientists believe that millions more species live on Earth today than they have discovered. Many of these live in rainforests. Unfortunately, Earth's rainforests are being destroyed. In fact, scientists think that Earth may be losing up to 137 plant, animal, and insect species each day because of rainforest destruction. That adds up to more than fifty thousand species every year.

This habitat destruction destroys many beautiful life-forms. It could also destroy the hope of finding cures for many different diseases. Scientists have already developed 120 drugs from various plants. These medications are sold around the world today. But researchers have only had a chance to test about 1 percent of the rainforest's plants. What if one of the fifty thousand species lost to rainforest destruction this year could have been the active ingredient for the cure of a deadly disease? If that plant disappears forever, humans will never know its potential benefit. And once a plant or animal species goes extinct, the possibility of it evolving into another species is gone.

While evolution, as a process, obviously predates modern humans, people have great influence on the evolution of other creatures, unintended or not. Humankind needs to take greater responsibility for their environmental impact and protect Earth's other organisms. We need each other more than we know.

Forests are home to approximately 80 percent of terrestrial biodiversity.

GLOSSARY

ADAPT To change and become better suited to an environment.

ALLELE A different version of the same gene.

AUTOSOME A chromosome that does not determine sex.

BIPEDAL Walking on two feet.

CHROMOSOME A structure containing genes that is passed down from parent to offspring.

CREATIONISM The belief that matter, the forms of life, and the world were created by God out of nothing and often in the way described in the Bible's Book of Genesis.

DEOXYRIBONUCLEIC ACID (DNA) A large chemical molecule in nearly all cells that carries genetic information.

DOMINANT GENE A gene that will show its effects even if there is only one copy of the gene in the genome.

EVOLUTION The process by which populations change over time due to genetic alterations.

EXTINCT An organism that has died out or ceased to exist.

GAMETE A reproductive cell that has half the number of chromosomes as other cells.

GENE A sequence of base pairs that tells the body how to make a particular protein.

GENE FLOW The introduction of new genes into a population because of the movements of individuals from one group to another.

GENETIC DRIFT A situation in which the frequency of alleles in a population changes by chance.

GENETIC EQUILIBRIUM A situation in which no evolution is taking place.

GENOME All the genetic information in an organism.

GENOTYPE The genetic makeup of an organism.

GEOGRAPHIC SPECIATION A situation in which a new species is formed because of a physical separation between a small group of organisms and the main population.

HABITAT The natural home or environment of a plant or animal.

HOMINID The group of bipedal primates that includes apes and humans.

HOMOLOGOUS Describing a structure with similar anatomy but different functions.

INTELLIGENT DESIGN The belief that matter, all forms of life, and Earth were created by a designing intelligence.

MACROEVOLUTION Evolution at the species level that occurs due to genetic change over generations.

MEIOSIS Cell division that reduces the number of chromosomes in half and produces gametes.

MICROEVOLUTION Smaller evolutionary change that occurs due to genetic change over a relatively short time period.

MUTATION A permanent change in the order of base pairs in the DNA molecule.

NATURAL SELECTION The process by which organisms best adapted to their environment survive to reproduce.

PHENOTYPE Traits that can be seen or otherwise detected.

RECESSIVE GENE A gene that will only show its effects if there are two copies of the gene in the genome.

RECOMBINATION The process of gene swapping that occurs during meiosis.

SCIENTIFIC THEORY A well-studied explanation for observations and experimental data.

TRAIT A genetically determined characteristic.

FOR MORE INFORMATION

AMERICAN MUSEUM OF NATURAL HISTORY
200 Central Park West
New York, NY 10024-5102
(212) 769-5100
Website: www.amnh.org
Founded in 1869, the American Museum of Natural History is a place for visitors to discover more about human cultures as well as the natural world.

AMERICAN SOCIETY OF HUMAN GENETICS
6120 Executive Boulevard, Suite 500
Rockville, MD 20852
(301) 634-7300
Website: www.ashg.org
The American Society of Human Genetics works to promote the benefits of human genetics and genomics research.

CANADIAN MUSEUM OF NATURE
240 McLeod Street
Ottawa, ON
Canada
(613) 566-4700
Website: nature.ca/en/home
Visitors will find plenty of fossils and other evolutionary exhibits at this preeminent Canadian museum.

CHARLES DARWIN FOUNDATION
Charles Darwin Research Station
Av. Charles Darwin s/n
Puerto Ayora, Galápagos
Ecuador
Website: www.darwinfoundation.org/en/
The Charles Darwin Foundation is a nonprofit organization dedicated to responsible scientific research on the Galápagos Islands since 1959.

FLORISSANT FOSSIL QUARRY
18101 Teller County Road 1
Florissant, CO 80816
(719) 748-3275
Website: www.florissantfossilquarry.com
This quarry is located on a site where, millions of years ago, volcanic ash trapped animals and plants in shale. There are still lots of fossils for amateur paleontologists to discover.

INSTITUTE OF GENETICS
Dalhousie University
1-A1, 5850 College Street
Halifax, NS B3H 4H7
Canada
(902) 494-6466
Website: cihr-irsc.gc.ca/e/13147.html
The Institute of Genetics supports human genome research related to health and disease.

THE LEAKEY FOUNDATION
1003B O'Reilly Avenue
San Francisco, CA 94129
(415) 561-4646
Website: leakeyfoundation.org
The Leakey Foundation provides funding for research that furthers knowledge of human prehistory.

FOR FURTHER READING

Atkins, Jill, ed. *Around the World in 80 Species: Exploring the Business of Extinction*. New York, NY: Routledge, 2019.

Cutter, Asher D. *A Primer of Molecular Population Genetics*. New York, NY: Oxford University Press, 2019.

Darwin, Charles. *On the Origin of Species by Means of Natural Selection*. New York, NY: SNOVA, 2019.

Folkersen, Lasse. *Understand Your DNA: A Guide*. Hackensack, NJ: World Scientific, 2019.

Fuller, Randall. *The Book That Changed America: How Darwin's Theory of Evolution Ignited a Nation*. New York, NY: Viking, 2017.

Henneberg, Susan. *Genetic Engineering*. New York, NY: Greenhaven Publishing, 2017.

Hoffecker, John F. *Modern Humans: Their African Origin and Global Dispersal*. New York, NY: Columbia University Press, 2017.

Keranen, Rachel. *Evolution*. New York, NY: Cavendish Square, 2017.

Lewis, Ricki. *Human Genetics: The Basics*. New York, NY: Routledge, 2017.

Neal, Dick. *Introduction to Population Biology*. New York, NY: Cambridge University Press, 2019.

Rauf, Don. *DNA, RNA, and the Inheritance of Traits*. New York, NY: Enslow Publishing, 2018.

Roberts, Alice, et al. *Evolution: The Human Story*. New York, NY: DK Publishing, 2018.

Smith, Charles H., et al. *An Alfred Russel Wallace Companion*. Chicago, IL: The University of Chicago Press, 2019.

Sneddon, Bradley. *Charles Darwin*. Broomall, PA: Mason Crest, 2019.

Wood, Bernard A. *Human Evolution: A Very Short Introduction*. New York, NY: Oxford University Press, 2019.

INDEX

S

T

U

V

W

Z

PHOTO CREDITS

Cover, pp.1–80 Who is Danny_Shutterstock.com; cover JuliusKielaitis/Shutterstock.com; p. 5 LukeandKarla.Travel/Shutterstock.com; p. 7 laikavoyaj/Shutterstock.com; p. 9 New Africa/Shutterstock.com; p. 11 Sashkin/Shutterstock.com; p. 12 Ody_Stocker/Shutterstock.com; p. 14 Rawpixel.com/Shutterstock.com; p. 17 Niyal/Shutterstock.com; p. 19 nobeastsofierce/Shutterstock.com; p. 20 https://commons.wikimedia.org/wiki/File:Charles_Robert_Darwin_by_John_Collier.jpg; p. 21 cynoclub/iStock/Getty Images; p. 22 Frank Bach/Shutterstock.com; p. 22 https://commons.wikimedia.org/wiki/File:Old_hinny_in_Oklahoma.jpg; p. 23 https://commons.wikimedia.org/wiki/File:Erasmus_Darwin,_after_Joseph_Wright.jpg; p. 25 https://commons.wikimedia.org/wiki/File:Title_Page_of_Lamarck,_%22Philosophie_Zoologique...,%22_Wellcome_L0033032.jpg; p. 26 https://en.wikipedia.org/wiki/File:PSM_V57_D097_Hms_beagle_in_the_straits_of_magellan.png#/media/File:PSM_V57_D097_Hms_beagle_in_the_straits_of_magellan.png; p. 29 https://commons.wikimedia.org/wiki/File:Editorial_cartoon_depicting_Charles_Darwin_as_an_ape_(1871).jpg; p. 31 https://en.wikipedia.org/wiki/File:Darwin%27s_finches_by_Gould.jpg; p. 33 https://commons.wikimedia.org/wiki/File:Ernst_Mayr_PLoS.jpg; p. 35 imaginariumau/Shutterstock.com; p. 37 Ian_Redding/Stock/Getty Images; p. 39 Sergey Podlesnov/Shutterstock.com; p. 39 Artush/Shutterstock.com; p. 41 Thiti Sukapan/Shutterstock.com; p. 43 blvdone/Shutterstock.com; p. 45 Marcio Jose Bastos Silva/Shutterstock.com; p. 47 Daniel Eskridge/Shutterstock.com; p. 49 gorosan/Shutterstock.com; p. 51 Amadeu Blasco/Shutterstock.com; p. 53 Nobumichi Tamura/Stocktrek Images/Getty Images; p. 54 Daniel Eskridge/Shutterstock.com; p. 56 https://commons.wikimedia.org/wiki/File:Rudimentary_hindlegs_spurs_in_Boa_constrictor_snake.jpg; p. 59 https://commons.wikimedia.org/wiki/File:Neanderthal_child_(1).jpg; p. 60 https://commons.wikimedia.org/wiki/File:Australopithecus_africanus_-_Cast_of_taung_child.jpg; p. 63 https://commons.wikimedia.org/wiki/File:Mary_Douglas_Nicol_Leakey_(1913-1996)_and_her_husband_Louis_Seymour_Bazett_Leakey_(1903-1972).jpg; p. 64 Juan Aunion/Shutterstock.com; p. 66 ducu59us/Shutterstock.com; p. 69 Ivy Yin/Shutterstock.com.